WINDSONGS

WINDSONGS
Poems about Weather

Douglas Florian

Beach Lane Books • New York London Toronto Sydney New Delhi

W
E
A
T
H
E
R

Weather

Weather describes our atmosphere.
Like whether it's cloudy,
or whether it's clear.
Whether it's **freezing** or frosty or **cool**.
Whether it's snowing . . .
Today there's no school!
Whether there's **mist**,
or **monsoons**,

or **t**
O
r
n
a
d
o
s.

Whether the sun will help
grow our tomatoes!

Atmosphere

The
atmosphere
that's here on Earth
is what we breathe in
from our birth.
It's mostly full of nitrogen
and just enough of oxygen.
There's lots of water vapor too,
though it's invisible to view.
The atmosphere that fills our sky
is only sixty miles high.
That isn't much
when seen from space—
a sheet of air that's held in place
by gravity so it won't fly
to Mars to fill the Martian sky
or Venus, where the atmosphere
stinks like rotten eggs,
I hear.

CO_2
H_2O
O_2
Ar
CO_2

SUN

Sun

The
sun is the force,
and the sun is the source
of whether it's raining or sunny, of course.
For heat from the sun sets the Earth's air in *motion*,
creating our weather while warming each ocean.
The tropics have hot air. The poles' air is freezing.
And temperate zones, in between, may be pleasing.
We're all of us children of sun, so to speak,
not solely on Sunday but **all**
of the week.

Cloud

I cloud the sun
to wreck your fun
when you go out to play.
I rain cascades
on your parades.
I love to ruin your day.
My body bars
your view of stars,
as well as an eclipse.
And I will haze
up sunny days
of your vacation trips.
To nip your nap
I thunderclap.
My voice is very loud.
I have no shame
about my name.
I'm **proud** to be a cloud.

CLOUD

Rain

First
one drop.
Then two.
Then four.
Rain is pouring
more and more
as buckets of rain
are flooding the floor.
I never saw it rain this hard—
I'm doing the backstroke
outside in my yard
!!!!!!!!!!!!!

Hurricane

It's reigning rain! A hurricane! Enormous storm in spiral form! With winds one hundred miles per hour that pack a punch of force and power! The seas all surge! The danger is high! Except for the calm within the storm's eye.

HURRICANE

H A I L !

Hail

Hailstones big as marbles.
Come outside and see!
Hailstones big as golf balls,
landing next to me!
Hailstones big as baseballs—
a good three inches wide!
Hailstones big as basketballs.
Let's scramble back inside!

Wind

I can't be captured.
Can't be held.
Can't be seen
and can't be smelled.
I can't be cornered.
Can't be kicked.
Can't be tasted.
Can't be licked.
I can't be pulled
or pushed around.
But I ***can*** *blow trees*
down to the ground.

FROST
N
E
W
S

Frost

Jack Frost stopped
by our yard last night,
and now the green leaves
sparkle white.
He turned all he touched
to ice and snow,
and made the fir trees
glisten and glow.
He painted patterns
on my windowpane,
then gently kissed
our weathervane.

Lightning

Lightning is frightening!
For me it's a scare!
A real stomach-tightening
charge in the air.
Lightning has bolts
of great volts
in its flashes.
When lightning
strikes,
I run
fifty-
yard
d
a
s
h
e
s
!

LIGHTNING

Thunder

Thunder breaks,
then thunder claps!
Thunder wakes
me up from naps!
Thunder rrroars!
And thunder rrrumbles!
Thunder awes!
And thunder humbles!
Thunder crackles!
Thunder crashes!
Thunder follows
lightning flashes.
Thunder drums
inside a cloud.
Oh, and,
by the way,
it's **LOUD!**

Snow

Snowflakes
Snowfalls
Snow lakes
Snowballs
Snowstorms
Snow swirls
Snowmen
Snow girls
Snow b l o w s
Snow shifts
Snowplows
Snow drifts
Snowshoes
Snow fun
Snow melts—
Snow's done.

S
N
O
W

TORNADO!
TWISTER

Tornado

I whirl and I swirl, and I twirl round the air!
I turn and I churn, and I yearn just to scare!
Grab hold of your hat,
and your cat, and your sister!
Get wind of my winds—
call me
Mister
T
w
i
s
t
e
r
!

Fog

The fog is just
a cloud that's lost.
A cloud that's gone astray.
It woke up in a hazy daze
and slowly lost its way.
It drifted down,
close to the ground,
then napped beside a hill.
And gave the day
ten shades of gray,
each un-fog-ettable.

F
O
G

F
L
O
1995
2007
2022
1997
1995
ELM
2015
2003
O
D

Flood

There used to be a street here
before there was a flood.
There used to be a house here,
but now there's only mud.
Our street looked like a river
that carried cars and trees,
a box, a boat—
such things that float—
and many memories.

Drought

No clouds in the sky.
The ground is dry.
No hope of precipitation.
The grass is brown.
The drought's in town.
And rain went on vacation.

D R O U
G H T

29 RAIN
30 FAIR
31 DRY
0 20 40 60 80 100

Weather Instruments

For temperature: thermometer,
air pressure: a barometer,
humidity: hygrometer,
wind speed: an anemometer.
Whether the weather is harsh or gentle,
measuring it is *instrumental!*

Meteorologist

I am a meteorologist:
the weather's my domain.
I always know
if we'll have snow
or sun or sleet or rain.
My forecasts are all perfect,
for I'm **never ever** wrong.

But today I called for sunshine,
yet it's hailing all day long!

METEOROLOGIST

D
E
W

Dew

Today at dawn,
when day was new,
upon the grass
I viewed the dew.
And do I love
the dew, the dew?
It's true!
I dew!
I dew! I dew!

Climate Change

When fossil fuels burn, they fill up the air
with carbon dioxide, which traps the heat there.
Then ice caps start melting and sea levels rise,
as strong greenhouse gases are stuck in our skies.
The oceans get warmer, then storms grow severe,
and floods and long droughts get worse every year.
Mars is too cold, and Venus too hot.
Our blue planet Earth is all that we've got.

EARTH
CO_2
N_2O
CH_4

Glossary

Atmosphere
The atmosphere is the blanket of gases that surround our planet. Without an atmosphere it would be too hot in the day and too cold at night for plants and animals to survive. It provides oxygen, which we need to breathe. It also protects us from harmful ultraviolet radiation from the sun. The earth's atmosphere is about 78 percent nitrogen, 21 percent oxygen, and 1 percent other gases, as well as water vapor. It is in constant motion, getting energy from the sun and from the oceans.

Climate Change
Climate is the long-term pattern of weather in an area. Climate usually takes many years to change, but scientists believe that recently there have been rapid changes to Earth's climate due to human activity. When we burn fossil fuels such as coal, oil, or natural gas for energy, carbon dioxide is released into the atmosphere. The carbon dioxide traps heat and raises the temperature of the air and oceans. This leads to more extreme weather such as heat waves, floods, or droughts. Warmer sea temperatures create stronger and more frequent hurricanes. As polar ice melts, sea levels rise, and animals lose habitats. Global warming is a danger to life on our planet. But there are solutions. We need to use less fossil fuels and more renewable energy such as solar, wind, and waterpower. We can remove carbon dioxide from the atmosphere by planting trees, which take carbon dioxide from the air. And we can reduce our use of plastics, which are made from oil and don't break down quickly in nature.

Cloud
Heat from the sun causes moisture from the oceans to rise in the sky as water vapor. If it rises high enough, it cools and changes into water droplets, forming clouds. There are three main types of clouds: cumulus, stratus, and cirrus. Cumulus clouds look like balls of cotton. They may grow bigger and darker to form cumulonimbus clouds, which bring rain. Cirrus clouds are thin and wispy and are found at high altitudes. Stratus clouds stay close to the ground and may come with drizzle.

Dew
If you see tiny drops of water clinging to plants and grass in the morning, even though it didn't rain, that moisture is called dew. It forms in the night when warm, moist air touches cooler surfaces such as flowers, leaves, or grass. The moisture from dew can help thirsty plants, but if they become too wet, harmful fungi may form.

Drought
A drought is a longer-than-normal period of dry weather. The lack of rain or snow dries out the soil and diminishes the water in streams, rivers, and lakes. Drought can damage crops and cause a water shortage. A drought can last for weeks, months, or even years and may lead to a shortage of food or even famine. Because of climate change, droughts have become more frequent and severe in some parts of the world.

Flood
A flood is an overflow of water onto normally dry land. A flood can be caused by a heavy rain, a coastal storm, or a broken dam. A flash flood happens when a sudden heavy rain falls faster than the ground can absorb it. It is dangerous because water levels can rise very rapidly without much warning. Cities may be at risk for floods because the sidewalks and streets can't absorb the water. Areas near bodies of water, such as rivers, lakes, or an ocean, are also at risk of flooding.

Fog
Fog is like a cloud that hovers near the ground. It forms from water vapor in the air that cools into tiny water droplets, often at night. As the air warms in the morning, the fog may disappear. But if the temperature stays low, the fog may last all day. Fog makes it difficult for drivers to see.

Frost
When water vapor touches a very cold surface, it can freeze, creating frost, a thin layer of ice crystals. It may appear overnight on the leaves and branches of trees or on the ground. Frost is often white, because in its freezing process it may trap air and scatter light in all directions. It can appear in beautiful patterns on windows.

Hail
Hail is formed inside thunderstorms when raindrops rise in strong winds called updrafts to very cold areas and freeze, forming hailstones. As more liquid water freezes onto the hailstones, they get heavier until they finally fall down to earth. Hailstones can be as small as one quarter of an inch in diameter or as large as six inches. They can damage buildings, cars, and crops, and they can even hurt people.

Hurricane
Hurricanes are huge storms that can stretch up to six hundred miles across, with winds more than two hundred miles per hour. Hurricanes form near the equator, where the ocean waters are warm. When warm, moist air from the ocean rises, the air pressure beneath lowers. Then air from nearby pushes in and starts to swirl. As the rising water vapor cools off, it forms spinning clouds of wind and rain. Hurricanes can cause great damage to life and property, especially in coastal areas.

Lightning
Lightning is a huge spark of electricity in the air created by a thunderstorm. It can occur between clouds, from a cloud to the ground, or from a cloud into the air. Lightning can damage trees, buildings, and cars. Occasionally people are struck by lightning. That's why it's best to stay indoors during thunderstorms.

Meteorologist
Meteorologists are scientists who study the atmosphere to understand and predict the weather and climate. They use weather instruments and gather information from weather stations around the world. By studying weather patterns, they can predict weather events, like rain showers or fog, or even warn us of dangerous weather such as tornados or hurricanes.

Rain
Rain is water droplets that fall from clouds in the sky. It is part of the Earth's rain cycle. In the beginning, heat from the sun warms water on the surface of the Earth. This water rises as a gas called water vapor. As the water vapor rises in the air, it cools, turning back into a liquid called water droplets, which form clouds. As more droplets cling together, they get so heavy that they fall back to Earth as raindrops.

Snow
Snow begins in clouds when temperatures are below freezing. Water droplets in the cloud freeze into ice crystals and form snowflakes that fall to the ground. If the weather is very cold, a powdery snow falls. But if the temperature is near freezing, the snowflakes clump together in a wet snow. A heavy, windy snowstorm is called a blizzard. Snow has a white color because it reflects light.

Sun
The sun is a star in the center of our solar system of eight planets, including Earth. The heat of the sun keeps our atmosphere in motion. But the energy from the sun varies in different places around the world. Where the sun strikes the ground directly, its heat is stronger than where it strikes at an angle. Because the earth's surface is curved, the climate is hotter in the tropics, where the sun's rays are more direct. In the polar regions, where the rays hit the ground at an angle, the sun's strength is weaker. The temperate zones are between the polar and tropic zones. Temperatures are less extreme there.

Thunder
Thunder is the loud crashing sound in the sky that follows lightning. It is caused by the rapid expansion of air created by a lightning bolt. Thunder is heard after lightning is seen because sound travels slower than light. From nearby, thunder has a sudden loud cracking sound, but from a distance, thunder is a low-pitched rumble that lasts longer because of echoes. If you count the seconds between lightning and thunder and divide by five, that's the distance in miles to the lightning.

Tornado
A tornado is a rapidly spinning funnel of air that stretches from a dark thundercloud to the ground. A tornado starts in a thunderstorm when winds moving at different speeds start to spin. A funnel cloud forms, and if it touches down to the ground, it becomes a tornado. Tornado winds can speed up to 250 miles per hour. They can destroy buildings and bridges and even flip cars into the air.

Weather
Weather is the condition of the atmosphere and includes sunshine, rain, fog, or clouds. Weather is always changing, and it's created by temperature, wind, and water in the air. Weather occurs in the lowest part of the earth's atmosphere, the troposphere, because that's where water vapor is found. Without water vapor there can be no clouds or weather. While weather can change in minutes, hours, or days, climate is a weather pattern over a long period of time.

Weather Instruments
Weather scientists (meteorologists) have many instruments to give them information about the weather. An anemometer measures wind speed by counting how fast three cups rotate. A barometer uses a column of mercury to measure atmospheric pressure. An electric thermometer measures air temperature by exposing a metal probe to the air. A hygrometer measures humidity, the amount of water in the air.

Wind
Wind is the movement of air on the Earth's surface. Wind occurs because different places on the Earth's surface have different temperatures. Warm air rises because it weighs less than cold air. As the warm air rises, cooler air rushes in below it, causing winds to blow. A light wind is called a breeze. A gale is a strong wind measuring between 39 and 54 miles per hour. A hurricane has wind speeds over 74 miles per hour

Weather Websites for Kids

NASA Climate Kids: climatekids.nasa.gov/menu/weather-and-climate/

National Geographic Forces of Nature: nationalgeographic.org/forces-nature/

National Oceanic and Atmospheric Administration SciJinks: scijinks.gov

National Weather Service: weather.gov/education/

Nova Cloud Lab: pbs.org/wgbh/nova/labs/lab/cloud/

SkySci for Kids: scied.ucar.edu/kids

Snow Crystals: snowcrystals.com

Selected Sources and Further Reading

Anderson, Jay, Jay M. Pasachoff, and John A. Day. *Peterson Field Guide to Weather*. Boston: Mariner Books, 2021.

DK. *Eyewitness Weather*. DK Eyewitness. Rev ed. London: DK, 2022.

Furgang, Kathy. *Everything Weather: Facts, Photos, and Fun That Will Blow Your Mind!* Washington, DC: National Geographic Kids, 2012.

Gibbons, Gail. *Tornadoes!* Rev ed. New York: Holiday House, 2019.

Purtill, Sharon. *The Wonder of Thunder: Lessons from a Thunderstorm*. Windsor, Ontario: Dunhill Clare Publishing, 2021.

Snedeker, Joseph. *The Everything KIDS' Weather Book: From Tornadoes to Snowstorms, Puzzles, Games, and Facts That Make Weather for Kids Fun!* New York: Everything, 2012.

Taft, Jean. *Worm Weather*. Penguin Core Concepts. New York: Grosset & Dunlap, 2015.

To Pierson, a lover of weather

BEACH LANE BOOKS
An imprint of Simon & Schuster Children's Publishing Division
1230 Avenue of the Americas, New York, New York 10020

Simon & Schuster: Celebrating 100 Years of Publishing
in 2024
For information about special discounts for bulk purchases,
please contact Simon & Schuster Special Sales at 1-866-506-1949
or business@simonandschuster.com.
The Simon & Schuster Speakers Bureau can bring authors
to your live event. For more information or to book an event,
contact the Simon & Schuster Speakers Bureau at
1-866-248-3049 or visit our website at www.simonspeakers.com.
The text for this book was set in Neutraface Text.
The illustrations for this book were rendered with gouache paint,
colored pencils, and rubber stamps on primed paper bags.
Manufactured in China
0124 SCP
First Edition
2 4 6 8 10 9 7 5 3 1
Library of Congress Cataloging-in-Publication Data
Names: Florian, Douglas, author.
Title: Windsongs : poems about weather / Douglas Florian.
Description: First edition. | New York : Beach Lane Books, [2024]
| Audience: Ages 4-8 | Audience: Grades 2-3 | Summary:
"Discover the poetry of our weather in these poems about
everyday and once-in-a-lifetime natural phenomena"—
Provided by publisher.
Identifiers: LCCN 2023032041 (print) | LCCN 2023032042 (ebook) |
ISBN 9781665937726 (hardcover) |
ISBN 9781665937733 (ebook)
Subjects: LCSH: Weather—Juvenile poetry. | Meteorology—
Juvenile poetry.
Classification: LCC QC981.3 .F56 2024 (print) | LCC QC981.3
(ebook) | DDC 551.6—dc23/eng/20230816
LC record available at https://lccn.loc.gov/2023032041
LC ebook record available at https://lccn.loc.gov/2023032042